The John Danz Lectures

Of Men and Galaxies

By FRED HOYLE

University of Washington Press : Seattle

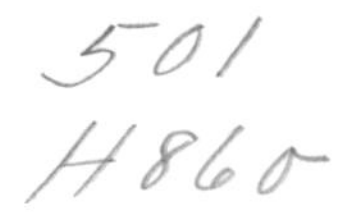

Library of Congress Catalog Card Number 64-25266
Printed in the United States of America

The John Danz Lectures

In October, 1961, Mr. John Danz, a Seattle pioneer, and his wife, Jessie Danz, made a substantial gift to the University of Washington to establish a perpetual fund to provide income to be used to bring to the University of Washington each year "... distinguished scholars of national and international reputation who have concerned themselves with the impact of science and philosophy on man's perception of a rational universe." The fund established by Mr. and Mrs. Danz is now known as the John Danz Fund, and the scholars brought to the University under its provisions are known as John Danz Lecturers or Professors.

Mr. Danz wisely left to the Board of Regents of the University of Washington the identification of the special fields in science, philosophy, and other disciplines in which lectureships may be established. His major concern and interest was that the fund would enable the University of Washington to bring to the campus some of the truly great scholars and thinkers of the world.

Mr. Danz authorized the Regents to expend a portion of the income from the fund to purchase special collections of books, documents, and other scholarly materials needed to reinforce the effectiveness of the extraordinary lectureships and professorships. The terms of the gift also provided for the publication and dissemination, when this seems appropriate, of the lectures given by the John Danz Lecturers.

Through this book, therefore, the third John Danz Lecturer speaks to the people and scholars of the world, as he has spoken to his audiences at the University of Washington and in the Pacific Northwest community.

Contents

Of Men and Galaxies

Motives and Aims of the Scientist

I AM NOT GOING to say very much about the more obvious questions that might be asked on this subject, but I will say a little to begin with. When I was recently in San Francisco I heard that an opinion poll had shown that a surprisingly large proportion of people were disturbed and suspicious about the scientist and his activities. What they undoubtedly had in mind was the nuclear bomb, the grisly horror that scientists are supposed to have unleashed on the world.

It is not usual for a really creative scientist to produce a weapon of war of his own volition. It is recorded that when the Greek colony of Syracuse was invaded by the Romans the great scientist Archimedes was persuaded to invent weapons the like of which had not been seen before—catapults, levers, cranes, all on an unprecedentedly big scale. Those Roman ships that were unwise enough to venture close inshore were seized by clawlike devices, bombarded by huge stones, overturned, and sunk. The defenders, greatly delighted with their victory, went on a three-day orgy. Meanwhile, the Romans landed down the coast, made their way overland, and finally entered the city from behind. In the ensuing sack, Archimedes was put to the sword because he was engaged on a geometrical problem, it is said, and refused to speak politely to the Romans when they arrived.

The essential point in this old story is that Archimedes was *persuaded* to make his cranes and levers, the patriotic motive being presumably the means of persuasion. It was the same with the nuclear bomb. Scientists were asked by their governments in time of war to produce the bomb. It is to be emphasized that work on the bomb was first started in Britain and was later carried to a conclusion here in the United States, both countries with democratic governments elected by the people. In short, the bomb was produced in our name. If we do not like it, we must blame ourselves, not the scientists who made it.

A word about the dictatorships, Russian and German. The dictatorships did not produce the bomb. There was no moral decision in this, however. The dictatorships did not produce the bomb because the people of a dictatorship are inherently less ingenious than the people of a democracy. Everything radically new is always produced in a democracy. In this connection the first sputnik was not conceptually new in any intricate way, not like the digital computer for example. The principle of the rocket was centuries old, and most of the development work necessary for the launching of satellites was done during the war and in military programs after the war. So I would not regard the Russian launching of the first sputnik in October, 1957, as an exception to my statement that dictatorships will always be behind in really new technological developments. The advantage of dictatorship is that coherent action is much easier to achieve—provided the dictator wants it. The disadvantage is that the natural inventiveness of the common people is suppressed. But this is by the way.

It is quite other matters that I wish to explore in this lecture. I want to inquire how far the aims, motives, and pressures to which scientists are now exposed are affecting science itself. I was born in the year 1915, exactly in the middle of the first thirty years of the present century. These first thirty years may be said to have been the great age of science.

I am not attempting to argue that remarkable discoveries were not made before the twentieth century or that remarkable discoveries are not being made now. The achievements of Isaac Newton in the seventeenth century, of James Clerk Maxwell in the nineteenth, as well as the recent discovery of the Omega-minus particle, with all its current significance, spring immediately to mind. But let me remind you of some of the discoveries of the first part of our century. First, Max Planck's discovery of the quantum and Albert Einstein's special theory of relativity: then Ernest Rutherford's discovery of the atomic nucleus, Niels Bohr's atomic theory, the general theory of relativity, and, not to overextend the list, the tremendous discovery of wave mechanics.

It would, I think, be correct to say that everything that has happened in the second thirty years of this century, at any rate so far as the physical sciences are concerned, was already foreshadowed in the first thirty years. On the practical side, atomic energy was all there, in the pioneer work of Francis William Aston in the early 1920's. This was also the source of a great deal that has happened in recent years in my own subject, astronomy and astrophysics. Our understanding of the processes that take place within stars, and indeed of the history of matter itself, owes its origin and inspiration to those two or three remarkable decades. Modern field theory, the heart of physics, spans the period. It began with Maxwell; its development has continued through the 1950's up to the present moment; but I do not think it can be denied that the biggest ideas belong to the period between Planck and Werner Heisenberg, to the first quarter of the century, in fact.

I would like to begin my inquiry into the aims and

motives of scientists by asking whether this was a matter of chance or whether something more was involved. We all know that the result of a baseball game, one particular game, may depend enormously on chance. The whole game can turn on whether a big hit goes fair or foul. Ten feet, depending perhaps on the way the ball curves in the wind, can make all the difference. Many games have to be played before it emerges which is the best club. The World Series is certainly not long enough to remove chance from the result, as any Yankee supporter with last year's Series in mind will tell you. In fact, a simple statistical analysis shows that about 150 games are necessary to get rid of chance effects, and this is exactly why the baseball season goes on for so long.

Chance effects also occur in science, and in all human activities. This, indeed, is one of the biggest problems the historian has to face up to, perhaps his biggest problem—to separate chance effects, *noise* as the physicist calls it, from systematic trends. I often suspect that some of my historical colleagues make the mistake of thinking it necessary to know everything about everything. Certainly teaching schedules and examination papers suggest all too often that history has become bogged down in fantastic and trivial detail, and that no serious attempt has been made to separate what is random from what is systematic. There are, of course, obvious dangers in making such a separation. It is all too easy to make a wrong separation and conveniently forget those details that point to the mistake. But exactly the same problem occurs in the sciences as well as in the humanities, and scientists have evolved methods for dealing with it.

Rather curiously, effectiveness in the separation of noise from systematic effects, the *signal* as we call it, seems to depend inversely on distance from the physical sciences. Serious arguments about signal and noise hardly ever arise in physics. As we move to the biological sciences the situation is often less clear. There have been famous arguments, such as the old question of whether acquired characteristics could be inherited. By and large these controversies have been satisfactorily and finally resolved, however. But passing from the biological sciences to the social sciences, and thence to the humanities, the picture gets less and less clear. This is why historical studies involve so much detail, and why the historian's examination papers look so horrible. He is not quite sure what is important and what is trivial, so everything has to go in.

The physical scientist cannot avoid sympathy with the historian when he comes to consider the problem of the first three decades of the present century. Were the achievements of these decades simply chance fluctuations, or was there a deeper reason for them? Evidently this question is of enormous importance, not only to the scientist himself, but to everybody. If something systematic has been involved, the inference is that science today is in decline, not disastrously or catastrophically, but sliding gently downhill.

One small indication that this might be so is that the question is never asked. In my experience most scientists take it for granted that all is well, that we are forging ahead as fast as it is possible to do, that, in Voltaire's words, "All is for the best in this best of all possible worlds." This, too, I would take as a small symptom of decline. Of course one can be too self-analytical. It does not pay to think too hard about the complicated mechanics of how to walk down a flight of steps; if you do you are likely to trip and break your neck. But the facts in this case are so striking that they should not be ignored.

Counsel for the defense would, I imagine, open its speech to the jury with the plea that today things are different and that they are different in a way that is no fault of the scientist in general or of the physicist in particular. The crucial facts on which the quantum theory was based came from a study of the spectrum lines emitted by excited atoms. All that was necessary to obtain these facts was to heat some gas to a few thousand degrees in the laboratory, to make it glow, and to analyze the light from the glow by passing it through a spectograph. The basic facts could be acquired for an expenditure of a few hundred dollars. Now, from a purely scientific point of view, the problem of the modern high-energy physicist is closely analogous to that of the physicist at the beginning of the century. What is wanted nowadays is a spectroscopy, not of atoms, but of the very particles that physics used to consider as basic. For it appears that even these particles possess excited states, analogous to the excited states of atoms. When a particle changes its state it emits a kind of radiation, analogous to the glow of light from a hot gas. Thus, conceptually, the problem is very similar to old-fashioned spectroscopy, but economically it is completely different. To carry out any worthwhile experiment today in this field of physics costs millions of dollars, not hundreds. And easily within the foreseeable future the subject will reach the stage when experiments will cost hundreds of millions of dollars. Quite apart from the economics of the situation, a huge piece of equipment takes longer to construct than simple equipment, so progress is not only vastly more costly but also much slower. The validity of these arguments is beyond question; the issue is whether they represent the whole truth.

Everybody would admit that two of the best three or four discoveries in physics of the last ten years have come from very simple equipment, a little more expensive perhaps than the equipment used at the beginning of the century, but of negligible cost by modern standards. I refer to Rudolf Mössbauer's discovery, and to the discovery of the nonconservation of parity in weak interactions. It might be argued that these discoveries were exceptional, and that it would be unwise to read too much into them. I am not so sure about this. Certainly, I do not think that physics could have gone in any other way but that in which it has gone, namely building the huge contraptions of which the most outstanding are at Brookhaven and Berkeley in this country and at CERN in Geneva. These machines themselves can do us no harm. It is what we ourselves believe, what postulates we come to accept without question, that may be harmful. One very common postulate is that to be effective science, at any rate physics, must be big. It is this dinosaur mentality that may wreak the damage. The cases I have cited, of the Mössbauer effect and of the nonconservation of parity, show that up to the present the dinosaur postulate has not been correct. My prediction is that in the decade ahead it will also not be correct. It will not be correct, partly because remarkable new discoveries achieved by quite simple means will, I am sure, come along—man's ingenuity always seems capable of turning up the unexpected—and partly because it is in the nature of dinosaurs to reach an end, to become extinct.

A simple extrapolation of the trend of the last ten or fifteen years shows that the building of bigger and yet bigger machines must reach an end, if only because this activity cannot consume more than the total energy of the whole human species. At first sight this might seem like a sad confession of failure. It might seem as if the human species in its inquiry into the nature of things must of necessity come up against an ultimate barrier. I suspect that a way round the barrier will be found and that it will come not at all from dinosaur thinking. During the last year, astrophysicists have become interested in the properties of matter under conditions that can never be simulated in the laboratory. By this I mean conditions of density, of the spacing between particles, that not even the biggest machines we can contemplate would be able to approach. This is not very surprising. What would be surprising would be if the opposite were true. Think of the whole universe as a laboratory. Is it likely that within that laboratory conditions will not be found that cannot be produced in our own local laboratories?

Here I am suggesting a new type of thinking, of looking for our facts, not by setting up a deliberate experiment, but by using the information that reaches us from outside the Earth. I am suggesting that the sum total of that information greatly exceeds what can be achieved by all the experiments that can be performed in a local laboratory. The importance of laboratory experiments lies, not in their breadth or wideness, but in just the opposite, their narrowness. By setting up special conditions, by *designing* an experiment, it is possible to simplify and to separate into many pieces problems that would be too complicated if we were obliged to work with the full wealth of detail that exists in the universal laboratory. My point, however, is that the very need for local laboratory experimentation is an indication of a lack of sophistication. Of necessity one has to begin by following the simplest path. But this does not mean that the simplest path must always be followed. I foresee a day, perhaps not very far distant, when more and more use will be made of the universal laboratory. There will, of course, be no sudden transition from one method to the other. When difficulties arise in the interpretation of the things we see in the universal laboratory, then we must simplify things by making local experiments. But there will be a shifting balance between the two methods, and the shift will move steadily toward the wider laboratory. One of the dangers of dinosaur thinking is that it cannot remotely conceive this possibility.

I want now to come back to the question, is there a systematic difference between the physics of today and that of the first quarter of the present century? My belief is that the answer should be, yes. I believe that the cultural pattern in science today is vastly different. Dinosaur activities are an obvious cause of the difference. The old free and easy conditions have gone, the sort of conditions I remember when as a student in Cambridge, thirty years ago, I used to tiptoe past Rutherford's laboratory. We now have vast, slick, streamlined laboratories, more reminiscent of an industrial production plant than of a laboratory in the old sense. On the face of it, things look more efficient. But are they? I have a personal theory that in order to be efficient about the things that really matter it is necessary to be inefficient about the things that do not matter so much. If one is obliged to be efficient about everything the best that can be achieved is a moderate measure of competence. To produce high peaks of inspiration it is necessary that there should be low troughs, implying, I believe, some degree of muddle and inefficiency. Such variations are essential, just as they are in a great musical work where a moment of tenseness and excitement is built out of periods of quietness and calm.

I am not alone in deploring this change. Remarks very much along the lines of what I have been saying were made to me only a few months ago by the director of one of the largest of these superlaboratories. But it is not easy to see how things can be changed. Facilities costing upward of ten million dollars cannot be treated in quite the same happy-go-lucky fashion in which our forefathers treated their laboratory equipment. By this I mean that laboratories cannot be shut down for four or five days while one goes off on a fishing trip in the mountains. The trouble is that it is the fishing trips that lead to the big ideas. For better or for worse we seem to be stuck with a new cultural pattern in science. Big science involves a change in the behavior of scientists, and this I do not think we can avoid. What we *can* do is to stop the big science mentality from overwhelming all our science. We can recognize the unfortunate necessity of some dinosaur activity. We can avoid, above all, the mistake of thinking that unless one is big one is negligible. Maxwell, Planck, Einstein, Rutherford—none of these men depended on big science, they depended on big ideas.

The danger lies in imagining that things have changed irrevocably. It is widely supposed, in my country and in universities in this country that do not command big facilities, that the situation is hopeless. Far from believing it hopeless, I would assert that there are probably as many as twenty really major discoveries in physics which are waiting around for somebody to pick up and which involve no major facility. I would suspect that to have a major facility would be an active handicap, since it is usually the case that the facility dictates the scientist's thoughts rather than the other way about. It is rather like making the mistake of having one's office in too perfect a building. People who work in marvelous buildings are dominated by those buildings, whereas it is the other way round for people who work in rabbit warrens. The builders of the great European medieval cathedrals knew this perfectly well. Walk into a big cathedral, and it wipes your brain clean of every thought. The same thing happens when you walk into these wonderful modern office blocks. The same thing happens all too easily in big science.

But it does not end here. Once you come to believe that bigness is everything, that the only way to get to the top of the heap is to be big, and since bigness costs a great deal of money, it is obvious that there will be an almighty scramble for money. It is also obvious that there will not be enough money going around to make everybody big. So a system of lobbying is bound to grow up. The lobbying will not be confined to ambitious, incompetent scientists. The best men will also be obliged to lobby, unless they are prepared to see those who are less able pass them in the race. It will be rare indeed that a considerable sum is voted to a man who has not invested a considerable amount of time and energy in pressing his own claims. It will be necessary for him to sit on numerous committees. In this country it will be necessary for him to employ a news bureau, or at any rate for his university to do so. The great men of the early part of this century did not give press conferences, nor did they spend 50 per cent or more of their time sitting on committees. Whether you agree with me or not about the present state of affairs, about its desirability or otherwise, I think you must agree that the cultural pattern is different from that which operated in the early years of the century.

So far, I have not blamed anyone for anything. And insofar as I am now going to blame someone, it will be the whole human species. There is not, and never has been, a human being who was capable of thinking straight, except by checking his thoughts against objective experience. I am not overlooking the mathematicians in making this statement. It is true that a mathematician may claim to have proved a result without appeal to objective experience and that all other mathematicians of his own day may agree with him. But experience shows that mathematicians of a succeeding generation will not agree. What constitutes proof in one generation is not the same thing as proof in another. In choosing mathematics I am concerned with the minimum of human emotion. In any thought process where emotions enter strongly it is enormously more difficult to avoid subjective rationalizations. For this reason it is well-nigh impossible for even the most responsible scientist to know exactly how far he should go in his requests for financial assistance, whether from state, government, or private sources.

By now I am building up a fairly considerable list of new cultural traits: dinosauric bigness, a submergence of inspiration in a humdrum kind of efficiency, endless committees, a tendency to grab every penny in sight. But more is still to come. There will be no disagreement that science has changed, and is changing, the world. In point of fact science has been changing the world since man discovered fire and made his first crude stone tools, only it was not called science then. The tempo has increased so rapidly in the last decade or two that everybody is now aware of the fact. In physics it never happens that one thing affects another without there being some sort of reverse action. The recoil of the rifle butt on your shoulder as the bullet is fired is a well-known example. Analogously, we cannot expect science to affect society without there being a reverse action. And once the connection becomes strong the reverse action will also be strong. This is one of the major ways, if not the major way, in which the cultural pattern in science has changed during the past thirty years, in which it is radically and completely different today from what it was in the early years of the century. Just as some degree of big science is unfortunately necessary, so some degree of connection between scientist and government is probably necessary. I would say that the future of science depends on how this relation turns out. What has happened to date is not particularly reassuring, as the opening remarks of this lecture already show.

A strong sign that all is not lost, however, comes from the remarkable fact that no outstanding scientist is to be found anywhere in any government of any country in the world. When you consider that almost every walk of life, scientists apart, is represented in the inner councils of government, this is indeed a remarkable fact. Some activities, economics and law particularly, breed politicians; science does not. I do not think the reason lies in political ineptness, in a lack of ability of scientists, some scientists, to fight an election successfully; the infighting that goes on in the scramble for funds would not disgrace the most adept politician. The reason lies elsewhere. Take an athlete out of training for six months, and he is no longer an athlete. It may be a little hard to say take a scientist, even a very good one, out of science for six months and he is no longer a scientist, but the statement is not a serious exaggeration. I think the absence of scientists from professional politics means that above all else scientists, essentially every one of them, want to remain what they are, scientists. This is the most heartening feature of the present state of affairs.

But it is inevitable that if scientists do not find their way into professional politics they will be formed into committees and asked to make recommendations to governments on a multitude of problems, where those problems are sensitively affected by scientific knowledge. Military defense is the obvious example, the example I began with at the outset. More and more in this century, political leaders have become constrained by weapons technology. Political decisions have turned on the availability or otherwise of technical devices. From both a moral and a political point of view it would have been better for Britain to declare war on Germany in 1938, at the time of the Munich debacle, than in 1939. I am convinced that the decision was delayed a year because of a lack of aircraft and of an adequate radar screen in 1938. In such circumstances it is quite inevitable that the scientist will find himself under heavy pressure to give a considerable amount of time to government problems.

There is a widespread belief that we all do our best work when we are young. But most of the evidence for this belief comes from modern times, because the expectation of life was much less in former generations than it is now, and people did not in general live very long anyway. So from the point of view of the present discussion the argument is circular and achieves nothing. We do know that in the arts, particularly music, the proposition is not true. In fact the reverse is true;

the best work is frequently produced in later years. My belief is that the relation between age and quality of work is very largely cultural. Under modern conditions it is almost impossible for an outstanding scientist to continue effectively in his work beyond the age of about forty-five because of the inroads on his time that will be made by nonresearch activities. The insidious presumption is that a good man can afford some loss of time, in terms of our athletic analogy that he does not need to keep completely in training. It is, I believe, this presumption that does the damage. It is a handicap from which our forefathers did not suffer. I would claim that we simply cannot afford to lose half the working life of almost all of our best people, for it is precisely the best who are called on the most often.

As if all this were not enough, we now have the two cultures. Sir Charles Snow is perfectly right in asserting the existence of two cultures. Where he is wrong, in my opinion, is in suggesting that the second culture, the scientific culture, is desirable. Snow himself is fascinated by the concept of power, more I think than any other novelist, past or present. I do not mean that Snow is concerned with power for himself, but that he finds fascinating exactly the type of situation I have been describing. He likes the thought of committees meeting, of men flying to Washington, if need be across the Atlantic. It is exactly the things I have just been deploring that constitute the second culture. It is not by chance that our awareness of this second culture is quite new. In the early years of the century the second culture did not exist. There was then no sensible difference between the inspirations of the scientist and those of the musician, the writer, and the artist. It is only with the development of the concept of the "corridors of power," to quote Snow himself, that the second culture has emerged. My belief is that this second culture needs watching, not nearly so much from the point of view of the humanities as from the point of view of science itself. I believe it is potentially far more dangerous to science than it is to the humanities.

The future—and here I trespass a little on the subject of my third lecture—will depend much more on the environment in which the scientist is called on to operate than it will on the scientist himself. It is a mistake to imagine that potentially great men are rare. It is the conditions that permit the promise of greatness to be fulfilled that are rare. It is a mistake to imagine that men like Shakespeare, Michelangelo, Beethoven, Newton, or Einstein were unique specimens of the human species. Individuals with their inherent capabilities are being born all the time, everywhere, in all communities. What is so difficult to achieve is the cultural background that permits potential greatness to be converted into actual greatness. At birth we all possess the ability to grow up in any community, to speak any language, to accept any social convention, however absurd. We are even equipped to live in caves under the primitive conditions of the Stone Age, if need be. But it is quite a different story by the time we reach the age of twenty. As the years pass we become more and more highly specialized to the particular community in which we happen to have been born, eventually reaching such a high degree of specialization—geographical, cultural, technical—that we would be most unhappy to be obliged to change to any other conditions.

The abilities with which we are born for the most part do not survive this tremendous change. Occasionally, however, the improbable happens; an individual's great natural talent may be perfectly matched to the requirements of the society in which he is brought up. It is then that the promise of greatness is fulfilled. Examples immediately spring to mind: the poetry of the Elizabethan dramatists, Florentine painting, Viennese music, and I would say the science of the first years of the present century. We are still living with the memories of yesterday. The conditions that brought our forefathers to greatness have not wholly disappeared from present-day society. All is not yet lost. The traditions are there, and the ability is there. What happens in the future depends, I believe, on the way our civilization develops in the years ahead. It is probably a mistake to be too pessimistic. Equally it is a mistake to suppose that everything is bound to continue as it was in the past. The days of greatness did not continue for the Elizabethan poets, or the Florentine painters, or the Viennese musicians. Their days of greatness vanished, not suddenly and dramatically, but gently, with a gradual slide downhill. What I am questioning in this lecture is whether we today are sliding gently downhill. I do not suppose that anybody can say yea or nay for sure. It is well-nigh impossible when one is in the midst of a development to know exactly how things are going to turn out. But I think we can be certain, in general terms, of what is needed to halt the downward slide, if indeed the slide is taking place.

I say in general terms, because I do not know at all how the details would work out. The general principle of the matter is this: that which a community really wants, it gets. I place emphasis on the word "really." Obviously there is a sense in which every community throughout the world wants good science. Good science guarantees good technology; good technology guarantees freedom from hunger, and guarantees the military efficiency of the community in question. But this type of want is a selfish one and will not do. In a similar sense every community wants great achievements in the humanities; but I think most writers, musicians, and artists would agree that modern industrial societies do not *really* want good literature, music, or art. The essential point is to want the product, whatever it may be, for its own sake. Money is certainly not the answer. A hundred and forty years ago you could have bought a Schubert song for a few pence. Today you will not get one whatever you are willing to offer. It is as well to face the fact that the creative spirit cannot be bought. It cannot be engendered by five-year plans. My suspicion is that a community that claims to want creative activity in one field but not in others does not really want anything. This I believe to be the devastating answer to Snow's two cultures. I suspect that if there are two cultures, as I agree today there are, then tomorrow there will be no culture at all. It seems to me inevitable that if the creative person on the humanities side finds himself out of tune with society, then so will a creative man on the scientific side.

I said that I could say nothing about details, but I will end this lecture with one detail. Beware of efficiency. Remember that Einstein was generally regarded as a vague, impractical man. Many scientists still think this. Yet the truth is that Einstein's calculations were anything but vague; they had a level of precision and exactness of thought which those who accuse him of being impractical are themselves quite incapable of attaining. Remember, too, that the girl Mozart really wanted to marry said after his death that she had turned him down because she thought he was a scatterbrain, and that he would never make good. It seems to be characteristic of all great work, in every field, that it arises spontaneously and unpretentiously, and that its creators wear a cloak of imprecision. Wordsworth had matters right when he spoke of Newton: "The index of his mind, voyaging strange seas of thought, alone." The man who voyages strange seas must of necessity be a little unsure of himself. It is the man with the flashy air of knowing everything, who is always on the ball, always with it, that we should beware of. It will not be very long now before his behavior can be imitated quite perfectly by a computer.

An Astronomer's View of Life

I WANT TO BEGIN by asking what we mean by life. In the old days, a hundred and fifty years ago, people had the comforting belief that the chemical substances that go to make life, whether vegetable or animal, were radically different from inorganic substances, such as common salt. Then it was discovered that some of the simplest organic molecules belonging to life could actually be made in the laboratory out of inorganic substances—so the idea had to be abandoned, at any rate in its simplest form.

Some overtones of this old point of view have persisted until quite recently. Very complex organic molecules, proteins and nucleic acids, have been thought of as somehow different—if only because of their enormous complexity. Let me put it this way: is it possible to build from quite simple chemicals a complete replica of a human cell with all its chromosomes, DNA, RNA, and so forth? I mean this in a literal sense, not a practical one, for such an enormously complex problem of synthesis is far beyond present-day laboratory techniques. If we answer that synthesis is not possible, then we are taking the same line that was taken a hundred and fifty years ago, a line of thought that has once been shown wrong. My strong belief is that a denial of the possibility would again be wrong.

If we once admit that a single human cell might be accurately synthesized, then it follows that many such cells could be made and could be fitted together into the same pattern we find in an actual living human. What would the result be? Would it be possible to detect any difference between a real human, born in the usual way, and an artificial creature built up out of simple chemicals—carbon, oxygen, nitrogen, phosphorus, and so forth? Suppose the artificial man were an exact, precise chemical copy of the real man; would there be a difference?

An important scientific comment can be made here. Chemical similarity may not be sufficient to guarantee similarity of behavior; the same kind of atom in the same arrangement need not behave identically, because the state of the atom is also relevant. In my first lecture I spoke of the fact that atoms can exist in many states, and that there can be jumps from one state to another. The same applies to molecules, in fact the situation is more complicated for molecules. Unless all the atoms and molecules were put together in the right states, an artificial man would in fact be different from a real man. It is even possible that the artificial man would not function at all but would be a meaningless piece of chemistry. If so, the whole problem of making an artificial living creature would be vastly more difficult than simply putting the right atoms together in the right order, and even this is way beyond present-day techniques. However, granted sufficient technical competence the problem might be tackled. An interesting point now arises. It might be the case that it was not merely difficult but impossible to get things exactly right. If one wishes to be completely thorough it is necessary to get all quantum effects right—to reproduce the wave function for the whole structure exactly—and this may not be possible. There might be some small differences that are unavoidable. These could lead to a critical difference of behavior. I do not think it would, but the possibility cannot be excluded.

You may think that this is not the right way to go about discussing the difference between life and non-life—that the concept of "soul" should be brought in right away. You might be inclined to suppose that certain arrangements of chemicals possess some new quality, outside themselves. In that case, suppose we go back to biological evolution, beginning with the first self-replicating cell, and ask exactly where in evolution this new quality first appeared. Did it start with the first cell, or the first primitive animals—does the jelly-fish have it? Or must we await the mammals—does the rat have it? Or is it special to the human? I suspect that the more precisely these questions are argued the less you will find of any special quality. And I think that what you will find will turn out to be quite different from what might be expected.

Suppose you were asked to arrange the animals in some kind of order, with humans at the top. Zoologists speak of "higher" and "lower" animals. The list might look something like this: man, ape, dog, larger mammals (e.g., elephant), smaller mammals (e.g., squirrel), birds, fish, insects, and so forth. I am not attempting to be very exact here; a rough and ready ordering is sufficient to bring out my point, namely, that our arrangement corresponds to the degree of complexity of nerve systems and brains. Here I come to a curiously ironical situation. None of us relishes being compared to a computer, but when called on to justify our superiority over other animals we immediately take refuge in our intelligence, in our capacity to make judgments, and so on—qualities that depend inevitably on our brain power; in short, on the quality of the computer with which we are endowed.

The reason why comparison with a man-made computer raises unpleasant associations is, of course, that man-made computers are relatively very simple structures, much inferior in most respects to ourselves. This may seem rather remarkable in view of the arithmetical powers of man-made computers. But the ability of the brain to analyze visual data is far more remarkable. A considerable financial fortune awaits anyone who can produce a convenient electronic device for reading ordinary print, a simple enough operation for us but very hard for an inorganic computer. Two points have to be noticed. First, the ability to multiply two large numbers in a small fraction of a second has never been of serious biological importance, and consequently we have not been selected for this ability. Second, there is no point in our manufacturing computers to do those things that we ourselves do superbly well.

Suppose now that in the future man-made computers become less and less simple. Will there be any hard and fast dividing line between such computers and the brains of animals—say the rat? My belief is that the assumption of an electronic dividing line will prove just as illusory as the attempt to find a chemical dividing line has been. And once there is no dividing line between an inorganic computer and a comparatively simple animal, then there is no dividing line between such a computer and ourselves. The problem becomes one of degree, not of kind. My point is that we ourselves are computers produced by the universe, by the long process of biological evolution, whereas the computers in our laboratory have been produced through us as an intermediary—but still by the universe. Objectionable as this conclusion will appear to many, I see little point in resisting it; it happens to be true! In growing up a child receives many rude shocks in its encounters with the world. Similarly, the human species as a whole must expect many shocks as it grows up. One of these I am convinced is that the phenomena of consciousness, of intelligence, independence, aesthetics, are going to come in ways that may seem strange to us. We must be prepared to find in the larger universe outside the Earth not only creatures very much like ourselves but widely different ways of doing things, even "inorganic" collections of matter endowed with a sense of "justice," for example.

It is, of course, much easier to think in terms of creatures like ourselves. Our imaginations are hardly adequate to go much beyond this. Fantasies soon reach the realm of science fiction. But, where conditions like those here on the Earth are concerned, the problems are astronomical, biological, and chemical. The biological and chemical problem may be quickly stated. Given not just one planet like the Earth but a large number with similar conditions of temperature, of illumination by a star like the Sun, and of chemistry—the same elements and compounds as were originally present at the Earth's surface—what would be the extent of the variety established after five thousand million years? This is a rather precise quantitative question of a kind that is often asked in physics. The concept is one of an *ensemble* of planets.

One can begin by asking why, if the initial state of

affairs on two planets was precisely identical, should there be any difference in their development, even over a time span as long as five thousand million years. The answer would be, only if some single quantum event were of decisive importance—and this appears improbable for so large a macroscopic system as a planet. The question is important, however, because no two planets could in practice be precisely alike. Temperature would not be precisely the same—only the same to within some reasonable margin, say a few degrees. The problem, then, is to decide whether such small differences would lead to substantial differences in evolutionary history. My suspicion is that they would not.

It is hard to deny that different ways of producing a self-replicating cell from an initial chemical "soup" may exist. But none can really be much more efficient than cells of terrestrial life forms. It is possible to look at genetics from the point of view of information theory. How much information is required to specify even the simplest single cell? More than is required for a nuclear reactor or an oil refinery, more perhaps than for a star. Yet all the information is contained in a cell no more than a few tens of millionths of an inch in size—the nucleic acid, or DNA as the familiar term has it. Obviously the storage of information is enormously efficient. Indeed, we can be sure that significantly more efficient systems are *not* chemically possible. Ours is quite certainly one of the best ways of doing the job. There may be other ways, but ours must represent one of the main streams of life. So I would be very sure that many of our set of planets would possess life forms closely analogous to those on the Earth.

So much for basic chemistry. What of the biological evolution? That is to say, what of the evolutionary process whereby cells become fitted together into complex multicelled plants and animals? When you regard the profusion of living forms on the Earth, you may feel inclined to suppose that the variety that might develop on our set of planets could be enormous. In one sense it might be; in another sense I think the possible forms would be rather restricted. Let me explain what I mean by both these statements.

In the sense in which we regard a bird as being widely different from an otter, there would be a wide difference. The flight requirements of the bird dictate the details of a bird's structure—hollow bones and so forth—while the swimming requirements of the otter dictate the otter's streamlined form. It is just because of streamlining that an otter or a seal can swim so much more expertly than you or I can, not because the otter or the seal possesses more efficient muscles or lungs. The requirements for the bird and the otter are different, and that is why the two creatures are *superficially* different. In a similar way, it is to be expected that creatures on another planet would be superficially different from those on the Earth.

However, an otter and a bird are really very similar

in the basic features of their construction. Take the eyes for instance. Both respond to much the same range of the electromagnetic spectrum. Both form an image on the retina. Both transmit nerve impulses from the retina to the brain. In both the information from the eyes is subject to data processing in the brain —and so on. We take all these enormous similarities for granted in our usual ways of thinking, and it comes as something of a shock to realize how inevitable an instrument like the eye really is. One simply cannot contemplate creatures that would not develop eyes in the normal course of biological evolution, provided that the planet in question possessed a broad similarity to our own. An atmosphere transparent to sunlight and a sun emitting roughly the same sort of light as our own, some general distribution of a land and sea— these are obvious examples of the similarities we might expect to find, not just in our hypothetical set of planets but in the actual universe.

During the past few years it has become possible to unravel some rather detailed clues concerning the birth of our own solar system, especially as it affected the material of the planets. These details begin to take on an appearance of inevitability. For instance, it appears that the element boron (so beloved of the Richfield Gas Company) was produced during the early history of the solar system, along with the neighboring elements, lithium and beryllium. It so happens that lithium is rather easy to detect in the light of distant stars; boron and beryllium are much harder to find. What transpires is that lithium is found in quite exceptional amounts in newly formed stars, particularly those like the Sun. I should remind you here that stars are forming all the time from the gas clouds that inhabit the Milky Way. In the light they emit, in the spectrum lines, we find the telltale lithium, indicating that the same kind of process that occurred while our planets were formed also occurs when other stars are formed. By implication, it seems pretty certain that planets must be built in other systems, just as they were here, although I must emphasize that this cannot be checked directly, because planets moving around distant stars cannot be seen directly, even with the biggest telescopes.

How often do we think that other planetary systems might have been formed? Well, as often as there are common or garden stars like the Sun, about 10^{11} of them in our Galaxy alone, about 10^{20} of them in all the galaxies that can be seen with a big telescope. Such huge numbers make nonsense of attempts to argue that conditions are special in our system. Perhaps the distance of the Earth from the Sun is especially favorable to life. Perhaps the size of the Earth is favorable, perhaps its geological and chemical histories, and so forth. But among 10^{20} examples there must be a vast number where similarly favorable conditions obtain. It is this vast number that constitutes the set of planets we have been speaking about.

Our inquiry into biological evolution suggested that instruments such as the eye should be widespread. So should the skeletal structures of all living creatures. There cannot be many efficient ways of making a joint, such as the knee or an elbow, and creatures without joints would surely find life tiresome in the extreme. My impression is that the full totality of universal life forms must look like a fantastic zoo—widely different in the sense of the bird and the otter but similar, even remarkably similar, in the basic ground plan.

As I said earlier, there may be quite different chemical ways of constructing living creatures. For example, it has been suggested that life might be based on the silicon atom instead of on the carbon atom. But, as I have already remarked, there cannot be many ways of storing genetic information with an efficiency equal to that of a terrestrial living cell. And if there were a dozen other ways of doing things as well, or even a million ways, which is unthinkable, there would still be plenty of planets among our 10^{20} that did things in the same way as they are done here. It is these in which we have a special interest.

What I have said about the inevitability of the eye as a universal possession by living creatures applies with even greater force to the brain. What I will call by the loose term *intelligence*—data processing and calculation if you prefer—is an obvious biological advantage. Indeed sense organs, even the simplest sense organs, demand some measure of data processing. A chemical system, or a mixture of a chemical system with crude electronics, obviates the need for a highly developed brain in comparatively simple creatures like insects. But no major development of intelligence is possible, I suspect, without a brain based on electronic pulses. Only in the crudest, saddest examples can we expect creatures without the sort of electronic computer that we ourselves possess.

Next, realize that a brain is a delicate instrument that must be encased in some kind of protective armor —bone perhaps. Realize also that eyes are best placed at maximum height above ground, to give the greatest range of vision; realize that eyes should be sited near the brain, so that it takes the least possible time for optical information to travel to the brain; and what have you? Inevitably, a head.

The argument can be followed further if you feel like it. I would suspect that no tool-making animal, in the sense of the archeologist, can develop without possessing some means of sensitive manipulation of inanimate objects, in our case the hand. Teeth are obviously no good. We can see that from the strenuous efforts of other terrestrial creatures, efforts and failures. And if something cannot be made to work here, the chances are against it elsewhere. So I expect something like the hand. Not necessarily a hand, but something capable of fulfilling a similar function.

Suppose we attempt to track down the possibilities a little more closely for the special case of highly intelligent creatures. For any large measure of development, intricate communication is essential. The power of the human does not lie solely in the superiority of the individual human brain over the brains of other animals. None of us operates as an individual really. Each of us is a conglomerate of many individuals. Every time you hear a new thought, every time you turn it over in your mind, you change yourself. Throughout our lives we are constantly writing on our brains the thoughts of others. There is no conceivable connection between us and the creatures we would have become if we had been isolated at birth from all human contact. If you could see yourself as you would have been, I assure you that you would receive a severe and unpleasant shock. Our special power is that each of us is really a multitude of persons; each of us receives, and each of us gives. These are the processes we call education, whether formal education as in a university course, or informal as in the bringing up of children in the home.

To return to my point, I do not think that any appreciable level of intellectual achievement can be reached by any creature, anywhere among our multitude of planets, except through this remarkable one-in-many process that we call education. And without an intricate system of communication I do not think education in a refined sense would be possible. I am not denying the possibility of a mother bear's teaching her cubs by grunts and cuffs of the paw; I am denying the possibility of learning the theory of relativity by such means.

Can one have a good sound system under water? I doubt it, in spite of claims that have been urged for the dolphin. Sound waves experience greater distortion in a liquid than in a gas, especially if the liquid is in constant motion like the sea. For this reason I suspect that highly intelligent creatures will always prove to be land animals, or at least amphibians. And I also suspect there will not be many cases of high intellects flying through the air—by natural means, that is. But this for a different reason.

From a strictly animal point of view a bird is an enormous success. The idea of being able to move around, the distinction between a plant and an animal, is to collect food over a wide area instead of being limited to what happens to be on the spot. A bird is the best "mover arounder"; its food collection area is enormous; for migratory birds the collection area encompasses a good slice of the whole planet. But to be able to fly, a bird must be light—I have already remarked on the hollow bones of a bird. A bird cannot afford a large brain, not simply because of the brain itself but because of the large supply of blood which a large brain needs, and which requires a big pumping mechanism, the heart and so forth. It follows that where intellect is concerned the bird is a dead end. If this were not so, walking animals would never have stood a chance on this planet in competition with birds. It happens that from an evolutionary point of view the bird is a dead end—in many ways a charming one, but a dead duck nonetheless. So it must be on all planets like the Earth. Only for a planet with much smaller gravity could things be different in this respect, and much smaller gravity implies a planet substantially different from the Earth.

To sum up, then, I expect that all highly intelligent creatures on planets like our own, and based on a similar chemical system, will be land animals as we are and will possess substantial similarities of construction —eyes, skeletons, heads, and so on.

The point that interests me about all this, and that has interested me for many years, is what are the chances of communication with intelligences on other planetary systems moving around other stars? It is to this question that I wish to direct the remainder of this lecture.

One's first thought is of space travel. But any reasonable assessment of the possibilities of travel to a distant planetary system soon shows up the difficulties as overwhelming. Everybody knows the almost absurd cost of the projected journey to the Moon. Very slight improvements of present-day rocket techniques, and only slight improvements are required for the lunar voyage, cost tens of billions of dollars. In contrast, let us consider a gross improvement, one far beyond all present-day concepts. Suppose we could achieve speeds ten times greater than at present. It would still take ten thousand years to reach even the nearest star.

A search for the nearest inhabited planet might well take a million years.

There is a good biological-style argument against deep space travel. If we could do it, so could somebody else. It is unlikely, fantastically unlikely, that we should happen to be the first in the field. Lots of others would have arrived here before now. Why would they have left? This planet is valuable real estate. In fact, we ourselves would never have evolved at all on the Earth, because the Earth would have been filled up from outside. The fact that this did not happen leads me to believe that space travel is not merely difficult but impossible, and this leads to a type of argument that I like to use, not merely in speculation, but even in serious scientific work. It might be construed as a religious type of argument, but I will not apologize for this.

My experience with scientific problems, in particular those in astronomy, is that where alternative possibilities exist it is never the possibility that leads to a dead end, the possibility that lacks interesting consequences, that turns out to be correct. It seems to be an overriding feature of all physical laws that they become more elegant, simpler in a way, as we get to know them better, but that their consequences become more varied and complex. Any proposal to change our formulation of a physical law which simplifies the statement of the the law but increases the complexity of its consequences has a high probability of being an improvement on the existing state of affairs. I shall make use of this concept in much of what I say in my third lecture.

For the moment we are concerned with space travel. If space travel through the Galaxy were possible, which I believe it is not, then all the planets of the Galaxy would come to be populated by the first few creatures to become capable of leaving their own systems and journeying through the depths of space. From the point of view of interest, would there be any particular merit in having the human species populate the whole of the Galaxy? Instead of there being some 10^{10} humans, as there will be on the Earth by the year A.D. 2000, the number might be increased to 10^{16} or 10^{17}. But is there any interest in sheer weight of numbers? I do not think so. It would make for a very dull situation. Far better to keep humans on this planet, and to keep other creatures on *their* planets. In this way there can be millions of different kinds of planets with different kinds of creatures, what I have called a fantastic zoo. This is the way I would design it if the choice were mine, and the real state of affairs is hardly likely to be less interesting and varied than the way we would arrange things if the choice were ours.

But now how about communication? Here there can hardly be any harm. In fact the development of intelligence is closely connected with communication. We have already seen, in the case of our own species, that we have got ahead through the absorption by individuals of the knowledge and culture of the whole species, not only through the ability of our individual brains to take in the thoughts of those around us, immediately today, but through absorbing thoughts first propounded by men long dead—the work of Newton, for example, in science, or Beethoven in music, or Isaiah in religion. Here lies the special facility of the human species. But how if the same thing could be done on a far vaster scale, by picking up the thoughts and concepts of other species? How if there is a stage of development, far beyond that which we have reached, in which different creatures from all over the Galaxy have come to pool their discoveries?

Here we have two alternatives. The first possibility is that, for purely technical reasons, communication from planetary system to planetary system may turn out to be impossible; the second, that communication may turn out to be technically feasible. Suppose we apply our criterion of which alternative is the more interesting, in the sense I described a little while ago. I think we will all agree that for communication to be possible within our universal zoo would be far more intriguing in its implications than the very dull situation in which all interchange was completely barred. My suspicion is that life has little meaning, but must be judged a mere cosmic fluke, unless this alternative is indeed the true one.

To come to the technicalities, things turn out just as we might expect; communication is almost certainly feasible. Our present-day techniques are not adequate, but we are within sight of what is required. Suppose we were to build an instrument of the type of a radio-telescope, but to be used for transmitting radio signals as well as receiving them, and suppose we could construct a big saucerlike metal mirror of the size of the new instrument in Puerto Rico, 1,000 feet in diameter, but with the steerable precision of the Australian 210-foot telescope. Using such an instrument it would, I think, be possible to transmit an intelligible message as far as the nearest star, some four light years away, probably even farther. In saying this I have in mind that our transmission would have to stand out not only against the background emission of the Sun, but against the general background of cosmic radio waves, which constitute the study of radioastronomers.

Suppose we could manage to do a little better than this, to send our messages ten times farther. This would reach the nearest thousand stars. And if we could manage one hundred times farther, which I do not regard as out of the question, we could reach the nearest million stars. Somewhere among these, I suspect, is the neighbor we are looking for. Beyond this we do not need to go, because like a link in a chain our neighbor could relay our messages on to the next fellow, and so on. Since all we should have to say, in the beginning, would be something like, "Boys, we are here," further refinements need not concern us at first.

The time required for an interchange of messages would be enormously less than the time required for a space voyage. The latter we saw might take a million years. In contrast, an exchange of messages could be achieved in a few centuries. Even this is clearly not a project for the impetuous. But why should it be? Why should the logistic situation be adjusted for the convenience of the individual; why should it be adjusted to satisfy your or my personal curiosity? The important thing is the long-range development of our mental processes over centuries and even millennia. Viewed against the time scale of the development of human culture, a few centuries is quite correct. What is needed are the big thoughts, not the daily baseball scores. My point is that an interchange of messages could influence the future development of human culture, and for this it is by no means necessary to gabble continuously across the interstellar spaces.

Let me say a few words about what has been said and done on this subject so far. That it might be possible one day to receive signals from space I proposed some six or seven years ago. Following my proposal, there was an attempt at the National Radioastronomy Observatory at Green Bank, West Virginia, to detect code signals from nearby stars. The antenna was not large enough for the project to be viable, however, nor in my opinion was the frequency used high enough, or the bandwidth small enough. This is not intended as a criticism of the radioastronomers concerned, but rather to emphasize that it would be a matter of considerable difficulty to hit things right even if anybody happened to be beaming a message in our direction. And why should anybody be doing so? The solar system has been a dead system for nearly five billion years. Are we seriously to suppose that signals have been hopefully transmitted in the direction of the Earth over this vast span of time, in spite of the fact that no response has ever been sent out from the Earth? If we suppose this to be so, then we are attributing an enormously higher moral sense to our neighbors than we ourselves possess. I do not think any present-day government will underwrite a real attempt at interstellar communication, just because the project would be unlikely to pay off for several centuries. Yet we are calmly imagining that somebody or other has been underwriting the project of sending signals to us, not just for centuries, but for billions of years. This is exactly what we are assuming if we expect to pick up signals the cheap and easy way, by simply pointing our radiotelescopes at the stars. And, as I have just said, even this has not been done correctly or thoroughly.

I do not know whether Edward Teller was correctly quoted when he was reported to have said, "If there is anybody up there, why haven't we heard from them?" This seems to me like a man bottled up in a New York penthouse who says, "If there is anybody out there on the streets, why haven't I heard from them?" It is more appropriate to ask what we could possibly have to tell anybody out there that would be of interest to them. I suspect very little in the first place, although down through the millennia we may come to make our contribution to what I like to think of as "galactic culture."

We have reached the stage where I may come at last to the speculation with which I wish to close this lecture. You are all familiar with an ordinary telephone directory. You want to speak to someone, you look up his number, and you dial the appropriate code. My speculation is that a similar situation exists, and has existed for billions of years, in the Galaxy. My speculation is that an interchange of messages is going on, on a vast scale, all the time, and that we are as unaware of it as a pygmy in the African forests is unaware of the radio messages that flash at the speed of light around the Earth. My guess is that there might be a million or more subscribers to the galactic directory. Our problem is to get our name into that directory. We may achieve it sometime in the next few centuries. When we do we shall have to begin at the bottom of the ladder. A new subscriber is by definition one of the weaker brethren, since his knowledge has only just become sufficient for him to make the grade. I suspect that the position of our species will be like that of a child on its first day at kindergarten.

But what a pity, you will say, that we can never see all those other planets, that we can never see our companions in the cosmic zoo. Why so? I might just as well say what a pity that people in Seattle never see a World Series game. Of course you can, on television.

Indeed, some people say that you see more of a game on television than an actual spectator does. Well, interchange of messages does not just mean talk, or mathematical symbols. It implies the ability to exchange television pictures. There is no reason why we should not see other planets just as effectively as we can see our own on Cinerama, for example. There is no reason why there should not be plenty of film stars among other members of the zoo, why we should not be privy to a showing of some kind of cosmic *Tom Jones.*

Extrapolations into the Future

It is curious how much attention we all pay to the immediate future and how little to the more distant future. Perhaps we feel we have no control over what may happen in the centuries ahead. Or perhaps we are afraid of what our own thoughts might reveal. What do you think the world will be like five hundred years from now?

The most obvious prognostication is that the population of the world is going to rise, probably by a considerable factor. This has several obvious consequences, for example that it is a good idea to buy real estate. There is of course the problem of where and when. A big acreage up by the North Pole might well be enormously valuable a thousand years from now, when the ration of space will be down to one square yard per person if present trends continue. But a thousand years is a little long to wait, even for an astronomer.

Accompanying the rise in population there will be a corresponding rise in all forms of social pressure. In the simplest physical forms, there will be more noise, more traffic on the roads, more of everything. There will be fewer and fewer places to escape to, not only in this country but everywhere throughout the world. What the United States does today, Europe does tomorrow; and what Europe does tomorrow the rest of the world will try to do next week. I know several dozen places where I can get away from it all. Ten years from now the number will have dwindled; by the turn of the century it is possible that none will be left.

This gloomy picture is only a part, the lesser part, of the story. The real problems are not so much physical as psychological. Social pressures of all kinds are sharply on the increase. The indication is that they will continue to increase. As society becomes more amorphous, and as life becomes more aimless, the search for status will become more acute. In an economically backward country a man or woman works for survival, in order to eat. Such simple motives breed a healthy state of mind. Then as things become more prosperous we work to acquire luxuries, which may take many forms according to our tastes—we may work to buy a pleasant home, or to acquire leisure, or to buy an airplane. Make our society still more prosperous, and luxuries no longer remain luxuries. By definition a luxury is something you can have only rarely, and at some sacrifice. Once you can afford everything, nothing seems really worthwhile any more. In the present-day world there still remains leisure. But leisure, solitude, absence of noise and racket are all on the way out. One thing remains, however—status. This I see as the chief pursuit of the prosperous, glutted, overcrowded communities of the future.

Already, the greatest single social force in the developed countries is status, or face. And perhaps the greatest political force in the world today, nationalism, is simply a group form of status. All this is going to get a whole lot more intense.

Paradoxically, perhaps, the desire for status has played an important and honorable role in former times. It arises from the praiseworthy motive of wanting to be thought well of by one's fellow men. In a limited group—tribe, village, or small town—the surest way to be well thought of is to serve the group in some way, however humble. A great deal has been written about the narrowness of the small community, much of it not very well judged in my opinion. It is true that the literary man, the intellectual, may find the atmosphere of such a community stifling to his inventive powers. He will feel the urge to be away to the big city. But to most people the small community supplies the readiest fulfillment of our natural craving for social approbation. This craving has played a completely necessary part in the development of civilization, from its first crude beginnings in village and tribal communities up to the present day. It still acts as the spur for much of the inventiveness of those who work creatively, whether in the humanities or the sciences. Unfortunately, however, what used to be capable of ready fulfillment for everyone is utterly barred in our great cities. It is difficult to imagine a greater social difference between living in a village and living in a large city. For myself, I have lived more than half my life in villages. One knows everybody. One can know everybody's business —and one does. There is no dirty linen waiting to be washed in a village; it has already been washed, many times over.

In the big city, on the other hand, one knows almost nobody. It has already been said many times that there is no place so lonely as a city; it just happens to be true. The city provides no opportunity to be noticed except to a favored few. The majority simply exist, faces in the street. The prediction we are making is that it is all going to get much worse—except for the entertainment industry, for which I prophesy the brightest conceivable future. With increasing free time to kill, increasing monotony, increasing prosperity, there seems to be no limit to the potential development of entertainment. Any new idea, provided only that it is different, will come to suffice. More and more, the professions will cross over into the entertainment field. Those of us who are not employed directly in industry will come to realize that what we are really in is "show biz." This is already happening in science—it is so with 95 per cent of the space program. Incredibly, even the mortician has set himself up as entertainer, as you can see from any Forest Lawn billboard. Entertainment and real estate are the two sound investments for the future.

Is all this inevitable, you will be inclined to ask. After all, we the people decide what is going to happen, at least we do in countries with democratically elected leadership. This is the question I have been leading up to. Is it really true that we could change things if we wanted to do so? Let us look at this question first from an immediate, practical point of view, then on a wider basis.

It would certainly make a welcome change in our political campaigns if one or both of the contending parties were to take a major stand on issues like these: increase of privacy, less crowding in cities, smaller communities, and, above all, less pressure in our everyday lives, particularly in industry. For myself, I would welcome an attempt to make life less tense, less deadly earnest, to enjoy more of what we have got, to be less anxious about the future. But I am not convinced that such a policy would win votes; it runs directly contrary to all present-day trends. *If* it would win votes then I am now making an original political statement, for so far as I am aware no party has ever conceived of fighting its opponents in these terms. But I suspect that awareness of the problem is sharply correlated with income bracket; the higher brackets see it more clearly than the lower, just because experience has taught them that money will not buy peace, calm, or serenity, whereas the lower brackets tend to blame all misfortunes on absence of money. And of course the main voting power is held by the middle to lower brackets. Hence it would be impossible to win an election on this theme —and this brings me to a point, an important one I believe, that I would like to mention as an aside.

Have you ever considered the question of what decides the issues on which elections in democratic countries are based? We tend to take it for granted that they constitute the most important current social and political problems, the questions, in fact, that are going to affect our future in the most important degree. But is this true? I do not think so. In my experience the really important issues are never a subject of argument at an election. I risk your displeasure by asserting that the communist-anticommunist theme, so easily whipped up in this country, is not going to be one of the issues that will affect the future of the world in any really marked degree. I see the rivalry between the communist and anticommunist blocs in the same terms as I see the phenomenon of "display" among birds. It is a kind of mannequin parade, a sort of international fashion show. The real issues, I believe, rest on the impossibility of a long-term favorable future for the human species if different parts of the Earth remain in grossly different stages of development. On a long-term basis it simply is not possible to contemplate a life of prosperity and luxury in a few favorable cases on the Earth existing permanently alongside poverty and starvation everywhere else. Sooner or later, standards of living work themselves to a pretty constant level, like water finding its own level.

The reason why the communist-anticommunist theme raises our emotions so easily is that the problem involved is simple, easily expressed in black and white terms, like cowboys and Indians. All issues of this kind tend to have wide appeal if the time happens to be right for them. The time is not right today for the religious issues that split our forefathers into bitter factions. Today these issues seem what they always were, merely trivial.

The main reason, I think, why the really important problems never seem to be considered seriously, why the trivial always seems to displace them, is that the important problems arc not easily understood, let alone solved. I think most of us know in our hearts that the problem of what to do about the world as a whole, the problem of what to do about the phenomenal rise of the world's population, really constitute the issues that will affect the future. But we do not know what to do about them, so we dismiss these questions from our minds and concentrate on something more immediate and obvious. Frequent discussions of problems you cannot solve tends to be frustrating and irritating, so we tend to react by turning on the television.

In science, we proceed in much the same way. We solve those problems which we are capable of solving, and these are not usually the ones we would like to solve. There is no point in bashing out your brains when you just do not have the right idea. Many people seem to think that to do scientific research you need only to be a good experimentalist or to know a lot of mathematics and then go into a corner and think, or into the laboratory and fiddle about with something. But other qualities are needed. Important among them is the judgment of what problems are ripe for solution: exactly when does it become profitable to look again over old ground, to rediscuss problems that once seemed too hard. Originality springs from good judgment in this essentially aesthetic respect. It is one of the signs of the inexperienced young scientist, and of the amateur, that they are never content with small problems, that they must only tackle fundamental issues. Most scientists who continue to do this go to the wall. Most amateurs end up as embittered cranks.

It would obviously be unfair for me as a scientist to blame the politicians for proceeding in exactly the same way as we ourselves do, namely, for achieving what can be achieved and leaving aside those problems that cannot be solved at the moment. The issue is not one of blaming anybody. The issue I am raising is whether we ourselves are consciously in any control at all of the future. If we are not able to attack the problems that quite evidently will decide the kind of world that will exist in the future, then can we in any real sense be said to be in charge of our own destiny? This is the critical question I have been leading up to in this first part of my lecture. Let me state my opinion on this. It is my belief that man is not in charge of his future, and that he never has been. In the rest of this lecture it will be my purpose to develop this theme.

A word first about orthodox biology. When Darwin and Wallace propounded the theory of evolution in the middle of the last century, it was explicitly stated that man is no longer subject to evolutionary processes. It was supposed that because of free will, because of our freedom of choice, we can decide our own future as we please. I believe this view, which has undoubtedly persisted to the present day, to be wrong. As I have already said, I do not think that we decide the future at all, and I do not think that we are in any way free from evolutionary processes. The mistake of nineteenth-century thinking was to view the process of evolution in too narrow a light. You will recall the general ideas of biological evolution, that it arises from competition in part between members of the same species and in part between different species. The ground rules for the competition are supposed to be set by the physical environment: temperature, rainfall, the nature of the terrain, mountains or plains, the condition of the soil, and so forth.

In man's early days, competition with other creatures must have been critical. But this phase of our development is now finished. Indeed, we lack practice and experience nowadays in dealing with primitive conditions. I am sure that, without modern weapons, I would make a very poor show of disputing the ownership of a cave with a bear, and in this I do not think that I stand alone. The last creature to compete with man was the mosquito. But even the mosquito has been subdued by attention to drainage and by chemical sprays.

Competition between ourselves, person against person, community against community, still persists, however; and it is as fierce as it ever was.

But the competition of man against man is not the simple process envisioned in biology. It is not a simple competition for a fixed amount of food determined by the physical environment, because the environment that determines our evolution is no longer essentially physical. Our environment is chiefly conditioned by the things we know and the things we believe. Morocco and California are bits of the Earth in very similar latitudes, both on the west coasts of continents with similar climates, and probably with rather similar natural resources. Yet their present development is wholly different, not so much because of different people even, but because of the different thoughts that exist in the minds of their inhabitants. This is the point I wish to emphasize. The most important factor in our environment is the state of our own minds.

It is well known that where the white man has

invaded a primitive culture the most destructive effects have come not from physical weapons but from ideas. Ideas are dangerous. The Holy Office knew this full well when it caused heretics to be burned in days gone by. Indeed, the concept of free speech only exists in our modern society because when you are inside a community you are conditioned by the conventions of the community to such a degree that it is very difficult to conceive of anything really destructive. It is only someone looking on from outside that can inject the dangerous thoughts. I do not doubt that it would be possible to inject ideas into the modern world that would utterly destroy us. I would like to give you an example, but fortunately I cannot do so. Perhaps it will suffice to mention the nuclear bomb. Imagine the effect on a reasonable advanced technological society, one that still does not possess the bomb, of making it aware of the possibility, of supplying sufficient details to enable the thing to be constructed. Twenty or thirty pages of information handed to any of the major world powers around the year 1925 would have been sufficient to change the course of world history. It is a strange thought, but I believe a correct one, that twenty or thirty pages of ideas and information would be capable of turning the present-day world upside down, or even destroying it. I have often tried to conceive of what those pages might contain, but of course I cannot do so because I am a prisoner of the present-day world, just as all of you are. We cannot think outside the particular patterns that our brains are conditioned to, or, to be more accurate, we can think only a very little way outside, and then only if we are very original.

It is this pattern in which we are all conditioned that constitutes the most essential feature of our environment. Plainly, technology is an important part of the pattern, which brings me to what I believe to be a popular misconception concerning the role of the scientist in modern life. I think many people are instinctively aware that the old-fashioned notion of governments being completely in control of the course of events is wrong. It is wrong because governments do not invent technology, for instance. But the notion that scientists are now in the driver's seat is even more wrong. It is true that scientists produce the technology, but they do so at the behest of society, and they exercise no control over it. Why do not scientists stop, you may ask. Because they are every bit as much prisoners of the complex of thoughts that constitutes the pattern of our minds as nonscientists are.

In principle, you might suppose that scientists could stop the development of technology, or at least direct its development, by agreeing together on some policy. At one time I used to believe this myself. Now I see it just could not work like that, not merely because scientists would never reach agreement, but for a deeper reason. A physical scientist, for example, is only good at physics so long as he sticks to physics. He ceases to be good at physics as soon as he becomes a sociologist. The situation is that scientists produce science, and you must not expect them to produce anything else. If you do not like what they produce, then have no science. This in fact was the decision made in Italy after the time of Galileo, the four hundredth anniversary of whose birth we celebrate this year. The decision plunged Italy into centuries of unnecessary poverty, and perhaps for this reason the experiment is not likely to be repeated.

Humanists have repeatedly asked why Galileo did not stand up for his belief that the Earth moves round the Sun. Why did he recant? Because he was a scientist, and a scientist has no motive to become a martyr. To suppose otherwise is to miss the whole point of the scientist's position, to do science and that alone. In point of fact Galileo made his greatest discoveries *after* his recantation. Martyrdom might have pleased the humanists, but a great step, for which the world had waited two thousand years, would have been lost.

To return to the point at issue, who then is in the driver's seat? If not governments, if not scientists, who? Nobody. We are traveling in a vehicle that guides itself, just as our species has arisen from an evolutionary process that guided itself throughout past ages. It is my belief that nothing has changed, we are still in the grip of natural processes, we are not in charge of our own destiny. This is my belief, and the big question I wish to ask in the remainder of this lecture is whether we can say anything about where this process is likely to take us. Where does the road ahead lie?

I am going to make one big hypothesis—a religious hypothesis—that the emergence of intelligent life is not a meaningless accident. But I am not going to follow orthodox religions by presuming that I know what the meaning is. Intelligent life is such a remarkable phenomenon to emerge out of the basic physical laws that some connection seems implied, i.e., some correlation between laws and consequences of the laws —what in common terms we would call a *plan*. Let us see how much of the plan we can discover.

Without some control over the movement of inanimate materials there can be no development of an intelligent animal. An important reason for the emergence of man lies in his upright posture, freeing the hands and enabling them to fashion inanimate materials into tools. Now, moving things around requires energy. When we move something by hand the energy comes from the food we eat. Further progress, beyond simple manual processes, depends on the controlled use of energy. The first civilizations were based on the use of heat to smelt metals. To begin with, natural organic materials, particularly wood from trees, gave an adequate energy supply. With natural fuels it was possible to lift civilization to the level of the Greeks and Romans. Beyond this it was impossible to go without coal and oil, the basic prime movers of modern society. It is a matter of irony that neither Greece nor Italy possesses coal, otherwise the course of history would have been quite different.

How much coal and oil can you expect to find on a planet, not just the Earth, but on any one of the multitude of planets I was speaking about in my second lecture? Not, I think, enormously more than on the Earth. Perhaps ten times more in favorable cases, but any planet too heavily loaded with such materials would tend to be unstable. Coal and oil are highly unstable materials, igniting easily. It follows that intelligent creatures, evolving on any planet, will run into the same problems of fuel exhaustion—exhaustion of coal and oil—that we shall encounter in the next century or two. We already know what will have to be done about that: we must change our energy source to nuclear reactions, uranium and thorium, or perhaps the fusion of deuterium. My first point is that the same situation will arise on all planets, there will be an exhaustion of the simpler forms of prime mover. Sophisticated techniques, nuclear reactors, will then be required to take over.

It has been a very long step from the first crude charcoal smelting to a nuclear reactor. We ourselves have managed this step in about six or seven thousand years. A major fraction of the way has indeed been achieved in two or three centuries. Could we have made our discoveries at a more leisurely pace than this? No, we could not. The pace of discovery had to be very fast, otherwise all the coal and oil would have been gone before the advent of nuclear physics. Only a precipitate development will suffice. My second point is that the emergence of high intelligence as a more or less permanent going concern demands an almost discontinuous uplift. Millions of years of primitive conditions may precede the uplift, and millions may follow it, but the uplift itself must be made quickly, because of fuel exhaustion.

A whole lot of consequences follow from this simple observation. It has often been said that, if the human species fails to make a go of it here on the Earth, some other species will take over the running. In the sense of developing intelligence this is not correct. We have, or soon will have, exhausted the necessary physical prerequisites so far as this planet is concerned. With coal gone, oil gone, high-grade metallic ores gone, no species however competent can make the long climb from primitive conditions to high-level technology. This is a one-shot affair. If we fail, this planetary system fails so far as intelligence is concerned. The same will be true of other planetary systems. On each of them there will be one chance, and one chance only.

These inevitable arguments lead to a predicament. In a sudden lift from primitive conditions to sophisticated ones there is no time for an inherent change to take place within ourselves. In my first lecture I referred to the enormous adaptability of a child at birth. A baby possesses the ability not only to fit into a modern community but to fit into a primitive one too, even into the Stone Age. Indeed, it might even be easier to fit into the simple nomadic existence of the Stone Age than it is to make adjustment to the stresses of modern life. The reason for this is, of course, that the human species has not changed its inherent makeup since the Stone Age, not to any really appreciable extent. Our special problem today is just this: we are essentially primitive creatures struggling desperately to adjust ourselves to a way of life that is alien to almost the whole of the past history of our species. When I say this you must understand that by past history I do not mean just the last few centuries or the last few thousand years, but man's long evolution over tens and even hundreds of thousands of years. It was this long evolution that determined our basic physical and psychological characteristics, not recent history, not the period since the Romans or Greeks, for instance.

The point I am making is that this difficulty is unavoidable. As we have seen, the transition from primitive to sophisticated technology must be made swiftly—the resource problem demands that this be so. Today we are living at a unique moment, neither in the long primitive era nor in the better adjusted prosperous future. It is our century, our millennium, that must perforce take the maximum strain, for it is our fate to live during the transitional phase. And because we live in this special phase we find social difficulties, pressures, situations that defy even the simplest logical processes. We find ourselves in no real contact with the forces that are shaping the future.

It seems to me more than a small measure of comfort that our situation cannot be unique. Unless, indeed, we are the only intelligent creatures in the universe, which for the reasons given in my second lecture I find it impossible to believe, a similar situation must, it seems to me, arise whenever intelligence emerges. There must be the same sudden rise from the primitive to the sophisticated, the same psychological stresses on those who happen to live in the transition phase. Unless there is no meaning to be attached to life, to intelligent life, the problems with which we are faced must be soluble, indeed they must have been solved many, many times by creatures who are ahead of us in the process. There is a high probability that a way exists through the apparently impenetrable jungle of the future. What this way may be we can only speculate, and it is to such speculations that I wish to give the last part of these lectures.

It is to be emphasized that because a way exists it does not follow that the route is easy. Every mountaineer knows that while the escape from a dangerous situation may be possible it can still be desperately difficult to achieve. The process of evolution that has led to our presence here, at this moment, is not a pretty one. We exist today because of a past in which our forebears suffered untold distress and anguish. I am not referring simply to our human ancestors but to the long chain of creatures that preceded the human; their sufferings were certainly untold because they could not speak—we need only think of the shriek of the dying animal in the jungle. My point is that we are still in the jungle and that our descendants may come to say the same thing about us, that they owe *their* existence to *our* shrieks. There is no reason in principle why the future will be any prettier than the past has been.

Of course we do possess a far greater measure of control over the physical environment, and even over ourselves, than our forebears ever did. The big question is whether we possess sufficient control. My own conviction is that today we are still being swept along by the tide of events, like a canoe in the rapids. But ideas change, and as we understand more, both about inanimate things and about ourselves, the whole culture itself alters. Possibilities arise that were not present before. It is symptomatic of the present day that we are becoming aware of the rapids, and this awareness is already a considerable advance on the smug self-satisfaction of the nineteenth century. We are casting around for new concepts, as I have been doing here, and as John Danz was doing when he founded these lectures. New concepts are like genetic mutations, and like genetic mutations most of them turn out badly. This, too, I am afraid, is the fate of most lectures. But without mutations there can be no evolution, and without new ideas we should indeed be doomed. Over the decades ahead it is likely that, quite apart from changes dictated by the pressure of events, there will be really major changes brought about by new ideas, by argument, and by discussion.

An interesting example is the world population problem, treated in detail by my distinguished predecessor in this lecture series, Sir Julian Huxley. Fifty years ago this problem was hardly ever discussed. Today it has become a major issue of argument and discussion. It remains to be seen whether this attention will change things or not, whether our *awareness* of a serious problem leads to its solution. I suspect it may. There seems some evidence that the intellectual pressure to which even the Roman Catholic Church is being exposed is causing the Church to modify its position. If this should happen the power of thought and rational argument would be manifest, and there would at least be hope that the future is subject to our influence, that what we think matters.

We are very conscious of the importance of teaching children that they should not always follow the simplest line in their personal lives. We send them to school instead of allowing them to play, and at a later age we encourage them to spend several years learning a trade or a profession instead of following the immediately easier path of taking a job. We are all aware that unpleasant necessities have entered and will continue to enter our own personal lives. The same thing, it seems to me, can be said for our whole species. If we try always for the easy way out of every problem, things may turn out really more difficult in the end.

It is easier to do nothing, or very little, about the underdeveloped countries, but if we follow such a course it is inevitable that there will be world-wide unrest for an indefinite time into the future. It is probably inevitable that the poor countries will outpopulate the rich countries and that all hope of stabilizing the world population will be gone. And if the world population is not stabilized, if it rises as far as it possibly can, consistent with the best technology, nothing but pain and grief will follow. The future will then indeed be based on our cries of agony. It can readily be shown that such a situation must lead to catastrophic instabilities, that the world population will be subject to wild fluctuations, high peaks of overpopulation followed by low troughs of underpopulation. I have discussed in some detail what the effects of such fluctuations are likely to be.* From the point of view of the distant future the outcome seems reasonably good; from our point of view it is not good. Here, then, we have an outstanding example of the difference between deliberately taking a difficult path and following the easy way. It used to be a common theme of movie makers, and still is in television, to show how the youth who insists on always having a good time goes from bad to worse, eventually ending up among criminals.

*St. John's Lecture, University of Hull, 1963.

I suspect the same thing for our whole species: if we insist on always following the easy path we could end up as a criminal species.

It remains for me to attempt a summing up of these lectures. At root they have been concerned with the issue of how "that out there" affects us down here. Perhaps because I am an astronomer I believe much more in a relationship between things down here on the Earth and things out there in space than scientists in other fields are likely to do. Even in physics, there seem to me to be strong arguments to indicate that terrestrial experience is conditioned by external influences. In my own work I have been recently led to a formulation of the relativity theory different from that of Einstein. The nature of these lectures forbids me to discuss technicalities, but I can easily describe the results. In the usual theory the gravitational pull of the Sun on the Earth is fixed and is quite independent of what goes on "out there." In the new theory this is not the case. The Sun, the Earth, and the planets would behave quite differently if by some magic the distant parts of the universe were removed. Suppose the distant parts were gradually removed. The Sun itself would grow brighter and brighter, and the Earth would spiral in toward it. Life here would be fried to a crisp. Fortunately, there is no possibility of carrying out this test of the new theory, since not even the tens of billions of dollars spent on the space program are adequate to move the universe. But the issue will undoubtedly be settled in other ways. If the new point of view turns out to be correct, as I believe it will, physics becomes altered. Local experiments no longer suffice to enable us to understand physics completely. We will no longer be able to seal ourselves off from external influences, and even the most ordinary everyday events will seem to be conditioned by what is "out there." In the particular theoretical development I have been mentioning, it turns out that by observing the distant parts of the universe it is possible to determine how hard you will hit the ground if you are unfortunate enough to fall over a cliff.

The new results arise because qualities that have been thought inherent in individual particles, the *mass* in this case, turn out to be determined from outside. I suspect that other qualities of particles, for example electric charge, will similarly turn out to be determined by the universe in the large. The outcome would be a fusion of physics and astronomy, a blurring of the lines that have demarcated the two subjects in the past.

I mention these technical matters because the psychological situation in physics bears a remarkable similarity to the situation in human affairs. Just as physics has regarded itself as a localized subject of study, entirely discoverable through experiments performed in a terrestrial laboratory, so we usually think of ourselves here on the Earth as a complete unit, unaffected by what is "out there." In my second lecture, and in the present one, I have been questioning the validity of this point of view. The probability of there being intelligent life "out there" is overwhelmingly high. Events here on the Earth seem to me to be part of a general pattern, not special to us at all. In my second lecture I pointed out the inevitability of such instruments as the eye and the brain. Here I have argued that the most remarkable phase in the history of our species, that in which we are now living, the upsurge from a primitive Stone Age to a sophisticated culture and technology, is not just a chance affair. Things must go this way. We are following an inevitable path, one that must have been followed many, many times on other planets. You will remember that I raised the question of communication with other planets. I should like to refer back to this possibility by way of ending these lectures.

You may have wondered what explicit use such communication would be to us. I think we now have an answer to this question. If in other places other species have already followed the difficult route ahead of us, then plainly it would be an enormous advantage to know exactly where the dangers lie. My suspicion is that ample information exists in what I might call a "galactic library" to show exactly what is going to happen to us if the world persists in following current policies. It will be known, for example, what policies lead to nuclear war and what policies avoid it. Acquisition of such information would lead to perhaps the most revolutionary step in human thinking.

From time immemorial man has looked up at the sky in wonder. He has invariably placed his gods there. Instinctively he may well have been correct. To paraphrase the well-known psalm I will end by saying, "I will lift up mine eyes to the sky from whence cometh my help. . . ."

It may prove to be so.

Fred Hoyle

Internationally recognized cosmologist and professor of astronomy, Fred Hoyle has held, since 1958, the Plumian Professorship of Astronomy and Experimental Philosophy at Cambridge University.

Born in Bingley, Yorkshire, England, in 1915, where he received his early education, he entered Emmanuel College, Cambridge, and in 1936 qualified as a Mayhew Prizeman. As a graduate student at Cambridge he won the Smith Prize and was selected as Senior Exhibitioner of the Royal Commission of the Exhibition of 1851. In 1939 he was elected a Fellow of St. John's College, Cambridge.

At the outbreak of World War II, Professor Hoyle suspended his studies and astronomical observations to assist the British Admiralty in developing radar. Upon his return to Cambridge he began his studies on the internal composition of stars with Raymond Arthur Lyttleton. These studies led him to propose, in the late 1940's, the steady state theory of the origin of the universe in which he suggested that the creation of the universe was a process of continuous growth.

In 1956 Hoyle was appointed to the staff of the Mt. Wilson and Palomar Observatories in California. In 1957 he was elected a Fellow of the Royal Society, to which, in June, 1964, he announced his new theory of gravity. He also serves as a member of the British National Committee on Space Research.

Professor Hoyle has been a leader in the popularization of science and has often addressed his writings to a lay audience. In addition to numerous articles in scientific and professional journals, he has published Some Recent Researches in Solar Physics, The Nature of the Universe, A Decade of Decision, Frontiers of Astronomy, Man and Materialism, Astronomy, *and several books of science fiction.*